Caribbean Flora

C. Dennis Adams

Nelson Caribbean

The photographs were taken by the author who is on the staff of the Department of Biological Studies in the University of the West Indies, St. Augustine, Trinidad. Before moving to Trinidad, Dr. Adams was for sixteen years on the staff of U.W.I., Jamaica.

Dr. Adams has taught systematics and ecology in West Africa and the West Indies. During the past twenty-five years he has carried out research on tropical flora and vegetation and is the author of numerous publications on these subjects. These include 'The Blue Mahoe and Other Bush—An Introduction to Plant Life in Jamaica' and 'Flowering Plants of Jamaica'.

Thomas Nelson and Sons Ltd
Nelson House Mayfield Road
Walton-on-Thames Surrey KT12 5PL

51 York Place
Edinburgh EH1 3JD

P.O. Box 18123
Nairobi Kenya

Yi Xiu Factory Building
Unit 05-06 5th Floor
65 Sims Avenue Singapore 1438

Thomas Nelson (Hong Kong) Ltd
Toppan Building 10/F
22a Westlands Road Quarry Bay Hong Kong

Thomas Nelson (Nigeria) Ltd
8 Ilupeju Bypass PMB 21303 Ikeja Lagos

First published 1976
Reprinted 1980, 1984
ISBN 0-17-566186-3
NCN 500-6625-3

Printed in Hong Kong

A Look at Plant Life in the Caribbean

In this small book a selection of photographs has been arranged in groups covering, first, the more natural plant communities—the forests and savannas. Human influence is featured in sections on cultivated, useful ornamental and less useful, mostly introduced, plants, expressing in their own particular ways some of the delights and cautions of the southern Caribbean.

Forests and Savannas

It is hard to believe that the earliest accounts of Barbados are of an island covered with forest, because more than 350 years of European occupation have changed it out of all recognition. This change began slowly, as the pioneer settlers of the 1630's struggled to feed themselves by replacing the forest with small cultivations of yam, plantain, cassava, and maize, brought from the Dutch colonies in Guiana. Later they were able to grow cotton and tobacco, and these became important exports.

With uniform warmth, moderate rainfall, ample sunshine, and low undulating landscape, the island was ideal for arable agriculture, especially for growing sugar cane. To-day, except for a few patches of woodland in isolated valleys, the scene is of open fields, crossed by low stone walls and winding lanes.

In contrast, most of the other islands of the southern Caribbean still have a fairly extensive coverage of trees. This is because, although they are all warm, they are more rugged and rainy than Barbados. As a rule, the larger and the higher the island, the wetter it is. Either some natural forest has survived, or agricultural interest has centred around the growing of tree crops such as coconuts, cocoa, coffee, fruit trees, or spices.

The present-day flora of Barbados, in terms of variety and botanical interest, is a somewhat impoverished one. Nonetheless, few species have become extinct, in spite of the intensive agricultural activity.

Barbados

A fishing village in the parish of St Joseph on the east coast

The Portuguese explorers had noticed a great number of native fig trees, *Ficus citrifolia*, with their thick growth of fine hanging roots, and had referred to them, according to the eighteenth century historian Griffith Hughes, as 'Las Barbadas'—the bearded ones. It is from this description that the island is reputed to have got its name. The first English settlers, in 1627, also alluded to 'the beards of the fig trees'. In other parts of the West Indies a beard-like growth from trees would more likely be due to some epiphytic plant, for example the bromeliad *Tillandsia usneoides*—the Spanish moss or Old Man's Beard. It is characteristic of moist tropical forests that trees have epiphytes on them, whether these be mosses, lichens, ferns, orchids or, in the Americas, bromeliads.

Dead swamp

A late stage in the natural evolution of mangrove

If this gives the impression that primeval Caribaea consisted of islands entirely covered by dense forest, it is misleading and a great over-simplification. The islands are of many sizes and shapes and have different geological origins, so their ecology includes a great range of distinct habitats. Differences depend on factors other than warmth and the moisture deriving from the trade winds and proximity to the sea. Low islands, like Curaçao and Aruba, have a very low rainfall and support, at best, a vegetation of spiny thickets and cacti; the lowest parts of western Trinidad, lying within the influence of the mouths of the mighty Orinoco river of north-eastern South America, support a vast mangrove swamp dominated by brackish water and saturation atmospheric humidity. The Caroni swamp is the home of many small, mud-loving animals, enjoying an amphibious life-style among the roots of the mangrove trees. These in turn provide food for numerous birds, of which the gregarious scarlet ibis is the most spectacular.

At the other extreme, the volcanic peaks of the Windward Islands rise to about 5000 feet (1525 m) in Martinique and Dominica so that, at the highest points, frost occurs at night and fierce winds reduce the natural vegetation to a low tangle of heath-like plants. Between these limits, on various aspects of slope, and under conditions of different combinations of light, soil and drainage, the native and introduced plants of the islands thrive or subsist. Some of them are helped by man, others suffer from man's depredations; all survive together to create the unique character of the island scene.

Dominica

Epiphytes and climbers in a forest profile

Forests of tall trees exist in intertropical latitudes where the rainfall of the driest month exceeds four inches (10 cm). With that amount of rain in a month, water lost by evaporation does not exceed the supply. As the soil does not then dry out, plants rarely experience severe water shortage. In theory, a warm place with a total annual rainfall of forty-eight inches (120 cm) could support tropical forest, but in practice, on account of seasonal fluctuations, a somewhat higher annual rainfall is required to maintain tropical forest in its most highly developed form. The rainfall in most parts of the Windward Islands and Trinidad is considerably higher than seventy inches (180 cm), increasing to over two hundred inches (510 cm) on the windward sides of the higher mountains. Barbados has an average of about sixty inches (150 cm), but the Grenadines and the Dutch islands have much less than that.

Dominica is a very rainy island and large areas of its original forest remain almost untouched. This is because the mountainsides are very steep, and many parts are inaccessible. Where forest is being cleared, it is possible to see the complex profile of the vegetation and to appreciate that the trees make a framework upon which a great many types of climber and epiphyte depend for support.